SuperScience

World of WOW

Amazing Science Articles & Hands-On Investigations!

SCHOLASTIC

New York • Toronto • London • Auckland • Sydney
Mexico City • New Delhi • Hong Kong • Buenos Aires

Editor: Maria L. Chang
Cover design: Tannaz Fassihi
Interior design: Linda Rubes

ISBN: 978-1-338-32985-8
Copyright © 2018 by Scholastic Inc.
All rights reserved.
Printed in Heshan, China 62.

1 2 3 4 5 6 7 8 9 10   62   23 22 21 20 19 18

Contents

Dear Parent

Maybe you picked up this book because your child has an innate love of science. Or maybe your child is naturally curious about the world, and you want to help him or her learn more about how things work. No matter what drew you to this book, we're sure you and your child will find something that will fill both of you with a sense of wonder.

This book features 15 compelling topics in life, earth, and physical science with articles and companion hands-on investigations that will guide your child to discover the world around him or her. Each article, written in easy-to-understand language, provides context for the two investigations that follow. These investigations use readily accessible materials that can easily be found at home or at a local grocery store. Your child should be able to do most of the experiments independently, although you might want to be available to offer help as needed. And just like a real scientist, your child can write or draw about his or her observations, predictions, results, and conclusions on the recording sheet that comes with each investigation. To help you support your child's learning, we've provided background information and hands-on hints at the back of this book, written just for you.

"There are no seven wonders of the world in the eyes of a child. There are seven million."
—Walt Streightiff

Help your child get the most out of the learning experience with these quick tips:

- Provide a comfortable and quiet place for your child to work.
- Have your child conduct just one investigation a day.
- Help your child gather all the materials he or she will need for the investigation.
- Encourage your child's efforts, praising his or her successes and offering help as needed.

Ready? Then let's get started on this journey of discovery into the World of WOW!

— **The Editors**

Animal Babies

Babies Change and Grow!

Read about animal babies.
Then try the science investigations.

Emperor Penguin

A baby penguin is called a **chick**. Its head is black and white. Its body is covered in fluffy gray feathers called **down**. As the chick grows, its head turns black. It grows new feathers on its body—black for its back and white for its belly!

Silver Leaf Monkey

This baby monkey has bright orange fur. That helps the adults see it in the dark forest. As it grows older, the monkey's fur will change color. It will turn dark gray—just like its parents' fur.

Warty Newt

This is a newt. It lives in a pond. A newt starts out as an egg in the water. When it hatches, it becomes a **larva**. The larva breathes underwater with **gills**, just like a fish. But a newt's gills are on top of its head! When the larva turns into an adult, it lives on land.

larva

adult

Investigation 1

An emperor penguin does not leave its egg in a nest. It carries its egg on its feet! What is it like to walk with an egg on your feet? Try it!

1. Gather the materials you will need.

2. Pour the rice into the zip-top bag. Press out the air. Seal the bag and roll it up. Put it inside a sock. Tuck in the top of the sock. This is your penguin egg.

3. A penguin father carries his egg on his feet. Put your egg on top of your feet. Carefully walk around the room. If the egg falls off your feet, stop. Pick it up and put it back on your feet.

4. What did you do to make sure the egg stayed on your feet? (Did you turn your feet a new way? Speed up or slow down? What else?) Record on the next page.

5. **Think:** Why do you think penguin fathers carry their eggs on their feet? How do they keep the eggs safe?

Materials

★ 1 cup rice

★ quart-sized zip-top plastic bag

★ sock

★ recording sheet (next page)

1. Do Steps 2 and 3 of the investigation. What happened?

_____

_____

_____

2. What did you do to make sure the egg stayed on your feet?

_____

_____

_____

_____

3. Why do you think penguin fathers carry their eggs on their feet? How do they keep the eggs safe?

_____

_____

_____

_____

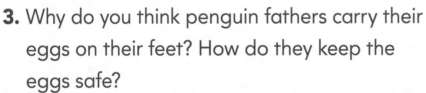

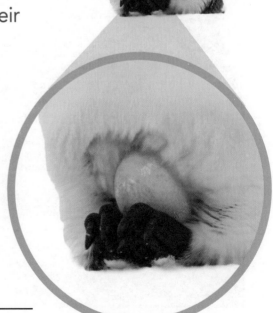

Investigation 2

Some baby animals have bright colors so their parents can find them easily. How can you make a baby bird stand out in its habitat (the nature around it)?

Materials
★ crayons
★ recording sheet (next page)

1. Gather your materials.

2. Look at the "pretend" habitat below. Look at the drawings of the baby birds on the next page. Imagine they will live in that habitat. Will the white birds stand out or blend in?

3. Color Bird 1 so that it will stand out in the habitat. **Remember:** You want your bird to be seen when it's in its habitat.

4. What if you don't want the baby to stand out? What if you want to hide it from enemies? Many baby animals have colors and patterns that make them hard to see. They blend in with their habitats. The name for this kind of blending in is **camouflage**. Color Bird 2 so it is camouflaged.

1. Look at the drawings of the baby birds below. Imagine they live in the habitat on the previous page. Do the white birds stand out or blend in?

_____

_____

2. Do Step 3 of the investigation. Is your baby bird easy or hard to see?

_____

_____

3. Do Step 4 of the investigation. How did you **camouflage** (blend in) your baby bird?

_____

_____

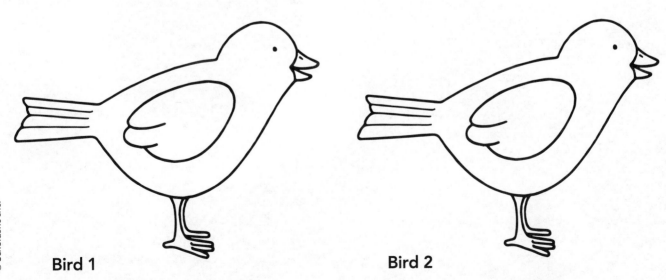

Bird 1   **Bird 2**

Plants

Is This a Plant?

Read about these amazing plants. Then try the science investigations.

The **giant sequoia** (seh-KOY-uh) tree is one of the tallest plants on Earth. Water travels from its roots all the way to the top. It takes almost a month!

giant sequoia

Most plants grow from seeds. In many plants, like the **sunflower**, seeds develop inside flowers. Animals, like birds, eat the seeds. They help spread seeds to new places to grow.

sunflower

Watch out, bugs! The **Venus flytrap** has a taste for insects. Most plants get food from soil. But this one traps bugs instead.

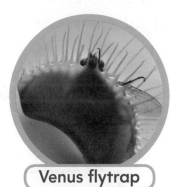
Venus flytrap

The **cactus** has thick, waxy leaves and stem. These store water. Its prickly spines keep away thirsty animals.

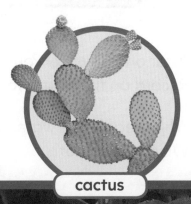
cactus

Investigation 1

In the hot, dry desert, plant leaves need to hold on to water. What kind of leaf stays wet in the desert? Find out here.

Materials

★ 3 moist paper towels

★ cookie sheet

★ wax paper

★ 2 paper clips

★ recording sheet (next page)

1. Gather the materials you will need.

2. Spread out one wet paper towel on the cookie sheet. This is Leaf 1.

3. Roll up the other two paper towels. Put one on the cookie sheet. This is Leaf 2.

4. Wrap the third paper towel in wax paper. Use a paper clip to keep each end closed. This is Leaf 3.

5. Put the cookie sheet in a warm, dry place. **Predict:** How will the leaves change after one day? Record your prediction on the next page.

6. After one day, feel your leaves. How have they changed? Record on the next page.

7. Which leaf keeps water in best? Which would make a good desert leaf? Which would not?

1. Do Steps 2–5 of the investigation. **Predict:** How will the leaves change after one day? Record your prediction in the chart below.

2. After one day, feel your leaves. How have they changed? Record below.

	My Prediction	What Happened
Leaf 1		
Leaf 2		
Leaf 3		

3. Which leaf keeps water in best? Which would make a good desert leaf? Which would not? Write your answers on a separate sheet of paper.

Investigation 2

How do seeds get to good growing places? Try this!

1. Gather the materials you will need.

2. **Think:** What's the first thing seeds need? A good growing place! But how do they get there? Read "How Seeds Go" below for some clues.

3. Look at your seed (popcorn kernel). How would you like it to travel? Will it glide through the air? Float on the water? Or will it stick to an animal passing by?

4. Test your seed. Can it travel the way you want? (Can it glide, float, or stick?) If not, change your seed. Add something to it. What things will you use?

5. Test your seed again. Did your seed pass its test? If not, make more changes.

6. When your seed passes the test, draw it on the next page. Tell how you changed your seed.

7. **Think:** Can your seed travel another way? How can you change it so it can?

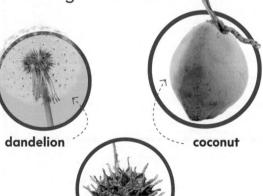

dandelion

coconut

burr

Materials

★ popcorn kernel

★ things to make your seed go (for example, tape, cotton, tissue paper, string, Velcro)

★ recording sheet (next page)

How Seeds Go

Some seeds glide.
Fluffy dandelion seeds are blown around by the wind. Will your seed glide through the air? Put it on your hand and blow. How far does it go?

Some seeds float.
Coconuts can float away in water. Will your seed float in water? Put it in a dishpan of water. Does it float or sink?

Some seeds stick.
Burrs are covered with little hooks that can hitch a ride on animals. Will your seed stick to an animal passing by? Press the seed onto a stuffed animal. How long does it stick?

1. Do Steps 2 and 3 of the investigation. How would your seed travel?

_____

2. Now do Step 4 of the investigation. How will you change your seed?

_____

3. Do Step 5 of the investigation. Tell what other changes you made.

_____

_____

4. When your seed passes its test, draw it in the box below.
Tell how you changed your seed.

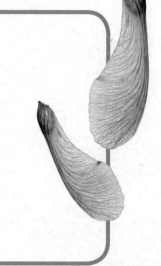

5. Think: Can your seed travel another way? How can you change
it so it can? Record your ideas on a separate sheet of paper.

Needs of Living Things

Saving Moto

Read about what a baby wildcat needs to survive on its own. Then try the science investigations.

A wildfire burned through grasslands in Africa. When the fire died out, some people found baby Moto. This little wildcat was all alone and covered in ashes. A wildlife photographer named Suzi took care of him. How did Suzi know what to do? She thought about wildcats and what they need to grow.

Suzi gave Moto milk from a bottle. After a few weeks, she started feeding him raw chicken. She also served him dead rats. That's what he would have eaten in the wild.

Then Suzi taught Moto how to hunt. Soon he could kill his own prey. That's important. Moto needed to do that so he could survive on his own. After six months, Moto went back into the wild.

Investigation 1

Dried yeast is actually a bunch of tiny living things. What does yeast need to live? Find out!

1. Gather the materials you will need.

2. Measure ¼ tsp of dried yeast into each of the four sandwich bags.

3. Press out most of the air and seal one bag. Label this bag "Control." Look at the yeast with a magnifying glass. In this form, yeast is not active. What do you think it needs to be active? (**Hint:** What do other living things need?)

4. To another bag, add 1 tsp flour. Press out most of the air and seal the bag. Label it "Food."

5. To another bag, add ¼ cup warm water. Carefully press out most of the air and seal the bag. Label it "Water."

6. To the last bag, add 1 tsp flour. Then add ¼ cup warm water. Press out most of the air and seal the bag. Label it "Food and Water."

7. Gently shake each bag to mix. Then set the bags next to one another on a table. Look at each bag closely. What do you observe? Record on the next page.

8. Wait 15 minutes, then observe again. What changes do you see? (**Hint:** When yeast "eats," it gives off gas. Do you see any gas bubbles?)

9. Wait 15 more minutes and observe again. What does yeast need to live?

Materials

★ measuring spoons and cups

★ 1 packet dried yeast

★ 4 zip-top sandwich bags

★ marker

★ magnifying glass

★ 2 tsps flour

★ warm water

★ clock or watch

★ recording sheet (next page)

1. Look at the yeast with a magnifying glass.
What do you think yeast needs to be active?

_____

_____

_____

2. Do Steps 2–9 of the investigation. Record your observations below.

Yeast	What I observed	After 15 minutes	After 30 minutes
Control			
Food			
Water			
Food and Water			

3. What does yeast need to live?

_____

Investigation 2

What foods work well for yeast? Try this!

1. Gather the materials you will need.

2. Different animals need different foods. (Could an otter eat the same food as a bee? Could a deer eat the same food as a shark?) What foods would you guess work well for yeast? Write your ideas on your recording sheet.

3. Choose two foods to test. Then make up an experiment to see which food works better. **Remember:** It's important to test only one change at a time. (You're changing the kind of food you add to the yeast.) Other things in your experiment should stay the same. That includes the amount of food you test. Write the steps of your experiment on your recording sheet.

4. How will you tell which food works better? (**Hint:** How were you able to tell the yeast was "eating" the flour?)

5. Do your experiment! Then tell what happened. Which food was better for your yeast?

Materials

★ foods you want to test (for example, milk, ketchup, peanut butter, sugar)

★ dried yeast

★ zip-top sandwich bags

★ measuring spoons

★ measuring cups

★ recording sheet (next page)

1. What foods do you think would work well for yeast?

_____

_____

2. Do Step 3 of the investigation. Write the steps of your experiment.

_____

_____

_____

_____

_____

3. How will you tell which food works better?

_____

_____

4. Which food was better for your yeast?

_____

_____

Habitats

A Home for Koalas

Read about koala habitats. Then try the science investigations.

Koalas live in Australia. But people are taking over their **habitats**, the places where they live. Now there's less food and shelter for the animals.

Koalas are very picky eaters. They eat only the leaves of eucalyptus (yoo-kuh-LIP-tiss) trees. People have cut down many eucalyptus forests. They have turned the land into farms, highways, malls, and houses. Today, there are not many koalas left.

Scientists and other people are now working together to help. They are planting new eucalyptus trees for the koalas. They hope this will help the animals thrive.

Investigation 1

What lives in your community park? Find out!

1. Close your eyes and imagine you are walking around your community park. What different ways do people use it? Record your thoughts on the next page.

2. What other living things do you think might use your community park? They might be living there year-round, or they might just visit. Make a list.

3. Go on a Community Park Safari. Look for signs of life. You can look for living things, like a tree or a butterfly. (Don't touch or disturb anything!) You can also look for clues that something was there. Maybe you will see a nibbled leaf, a spider web, or an animal track. Look carefully at what you find.

4. On your recording sheet, draw two things you found.

5. Share your discoveries with a friend. What surprised you the most?

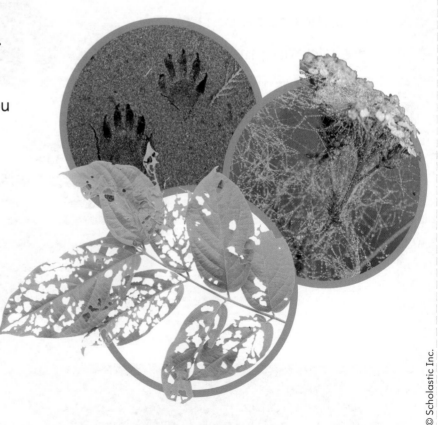

1. What different ways do people use your community park?

_____

_____

2. What other living things do you think might use your community park?

_____

_____

3. What signs of life did you find on your Community Park Safari? Draw two things you found.

4. What surprised you the most?

_____

_____

Investigation 2

What would you like to live in your community park? How would you change your park to make it a good home?

Materials
★ pencil
★ recording sheet (next page)

1. List three animals or plants you would like to have in your community park.

2. Look at your list. Does your community park have what each one needs to survive? (For clues, use the chart below.)

3. Pick one living thing that could not get what it needs from your community park. How could you change your community park so it would be a good habitat for your plant or animal? Write your ideas on your recording sheet. Draw a picture of the changes.

What Do Living Things Need?

Each kind of living thing has its own special needs. Here are some to think about.

Food	Some animals eat plants. Some eat other animals.
Water	Some animals just need water to drink. Some plants need to soak up water with their roots. But other plants and animals live in the water, so they need a lot more.
Shelter	An animal might need a cave or a good place to dig an underground burrow. It might need a branch and twigs for making a nest.
Territory	Some animals never travel very far. Some hunt over large areas. Others fly or swim thousands of miles every year.
Temperature	Some plants and animals need warmth. Some are happy in the cold.
Light	Some plants need a lot of sun. Some grow well in the shade.

1. List three animals or plants you would like to have in your community park.

_____

_____

2. Do Step 3 of the investigation. Write your ideas below.

_____

_____

3. Draw a picture of one or more of the changes below.

Pollinators

Way to "Bee" Helpful!

**Read about how bees help make new plants.
Then try the science investigations.**

1. A honeybee visits a flower to drink its nectar. Later, the bee will use the nectar to make honey.

2. The bee also gathers tiny grains called **pollen**. It takes the pollen back to the hive for food. Some extra pollen sticks to the bee's body.

3. The bee flies to a new flower.

4. Some of the extra pollen rubs off the bee's body. The pollen sticks to the new flower. Moving pollen from one flower to another is called **pollination**.

5. The pollen goes into the flower. It helps form seeds. A fruit grows around the seeds. Someday a hungry animal might eat the fruit. The seeds might drop to the ground and grow into new plants!

Investigation 1

How do bees pollinate flowers? Find out here!

1. Gather the materials you will need.

2. Make a paper-cup flower: Put 1 tsp of colored glitter in a small paper cup. The glitter is the flower's pollen. Make three more flowers. Use a different color of glitter for each flower.

3. Make a pipe cleaner bee as shown in the diagram below: Wrap a yellow then a black pipe cleaner around a pencil. Use another black pipe cleaner as a handle.

Materials

★ 4 tsps glitter (different colors)

★ 4 small paper cups

★ yellow and black pipe cleaners

★ pencil

★ recording sheet (next page)

4. Fly your bee into a flower. Gently tap your bee around the bottom of the flower. Your bee is drinking nectar and gathering pollen.

5. Fly your bee to a new flower. Choose one with a different color of glitter. Gently tap your bee around the bottom of that flower.

6. Repeat Step 5 with the other two flowers.

7. Return to each flower. Look inside. Do you see pollen from other flowers? (Look for glitter of other colors.) If so, your flowers were pollinated!

8. **Think:** The job of flowers is to attract pollinators. Why is that job important? Why are bees important to flowers? Why are flowers important to bees?

1. Do Steps 2–7 of the investigation. Do you see "pollen" from other flowers?

_____

2. Think: Why is it important for flowers to attract pollinators?

_____

_____

3. Why are bees important to flowers?

_____

_____

_____

4. Why are flowers important to bees?

_____

_____

_____

_____

Investigation 2

Different animals help move pollen from flower to flower. But each animal likes a different kind of flower. Design a flower to attract a pollinator.

1. Look at the four pollinators on the next page. Choose a pollinator and design a flower for that animal. Look at the list of things it likes. Draw a flower for your pollinator on the next page.

2. Gather materials and make a flower for your pollinator. Think about color. Think about shape. Think about smell.

3. Pick a different pollinator. Find a flower that it would like. Draw it on a separate sheet of paper. Write about which pollinator you think would like your flower and why.

Materials

★ things to make a flower (for example, colored paper, markers, crayons, scissors, tape)

★ a sweet-smelling liquid, like vanilla extract (optional)

★ a fruity-smelling liquid, like lemon extract (optional)

★ crayons or markers

★ recording sheet (next page)

What Pollinators Like

Bee
★ a place it can land (like a petal)

★ yellow or blue color

★ sweet smell

Butterfly
★ a place it can land (like a petal)

★ tiny tubes for butterfly tongues to gather nectar

★ pink or light purple color

★ sweet smell

Beetle
★ large flower

★ flat or dish-shaped flower (easy for beetle to climb on)

★ green or white color

★ spicy or fruity smell

Hummingbird
★ long tube shape (big enough for the bird's head and beak)

★ red or orange color

★ no smell needed

1. Do Step 1 of the investigation. Draw a flower for your pollinator.

2. Now do Step 3 of the investigation. Draw a flower for a different pollinator on a separate sheet of paper. Which pollinator do you think would like your flower? Why?

Weather

What's the Weather?

Read about the weather. Then try the science investigations.

A lot of what we do depends on the weather. We have to know how to get dressed for school, work, or even a picnic! If bad weather is coming, we want to know about it. We need to get ready!

How can we tell what the weather will be? One tool is a **thermometer**. It shows how hot or cold the air is. Another tool is a **wind sock**. It shows which way the wind blows.

The sky can also tell us the weather! Different **clouds** bring different kinds of weather.

Cumulus (KYOOM-yoo-lus)
If you see clouds like this, the weather will be nice.

Stratus (STRAY-tus)
If you see clouds like this, it might rain a little.

Cirrus (SIR-us)
If you see clouds like this, the weather is nice but may be changing.

Cumulonimbus (KYOOM-yoo-loh-NIM-bus)
If you see clouds like this, it might rain a lot!

Investigation 1

Scientists look for patterns in the weather. These patterns help them predict future weather.

Materials

★ pencil

★ recording sheet (next page)

1. Write today's date in the first blank column on the Cloud Tracker chart.

2. Look at the sky in the morning. Can you see any clouds?

   • **No:** Write "no clouds" on your chart.

   • **Yes:** Compare the clouds to the pictures below your Cloud Tracker chart. If they match one of the clouds, write down what kind. If not, describe the clouds or write "unknown."

3. Notice what the weather is like in late afternoon. Is there rain or snow falling? If so, is there a lot or just a little? Is it windy? Record the weather on your chart.

4. Check the weather again at bedtime. Record it on your chart.

5. Do Steps 1–4 on three different days. (Try to check the clouds at the same time each day.)

6. Look at your Cloud Tracker chart. Do you see any patterns?

Cloud Tracker

Date				
Kinds of clouds in the morning				
Weather in late afternoon				
Weather at bedtime				

Common Clouds

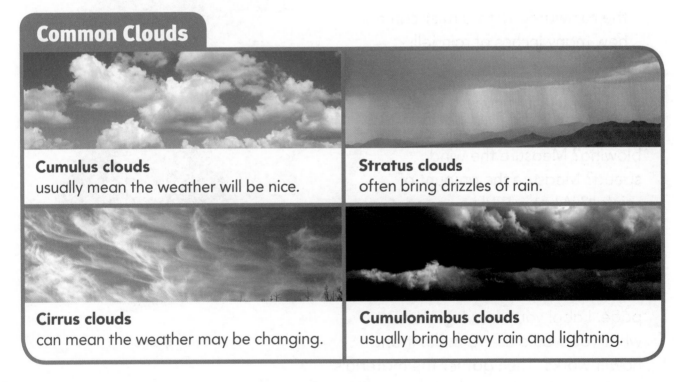

Cumulus clouds usually mean the weather will be nice.

Stratus clouds often bring drizzles of rain.

Cirrus clouds can mean the weather may be changing.

Cumulonimbus clouds usually bring heavy rain and lightning.

Look at your chart. Do you see any patterns in the weather?

Write your answer on a separate sheet of paper.

Investigation 2

Weather experts use tools to learn about the weather. Make your own weather tool!

1. Read about these three weather tools. Pick one you would like to make.

   - A **wind vane** shows which way the wind is blowing. The arrow points to where the wind is coming from.

   - An **anemometer** (an-i-MOM-uh-tur) measures wind speed. It uses three or four small cups. When the wind blows, it makes the cups spin.

   - A **rain gauge** (GAYJ) measures how much rain falls. A container collects the rainwater. Then a ruler can tell how many inches of rain fell.

2. **Think:** What weather tool would you like to make? Do you want to see which way the wind is blowing? Measure the wind speed? Measure the amount of rainfall? What would you need to make your weather tool? How would it work?

3. Draw your weather tool on the next page. Label your drawing. Write what you would use to make it and how it works. Then gather the materials you will need to make the tool. Finally, make your tool and try it out.

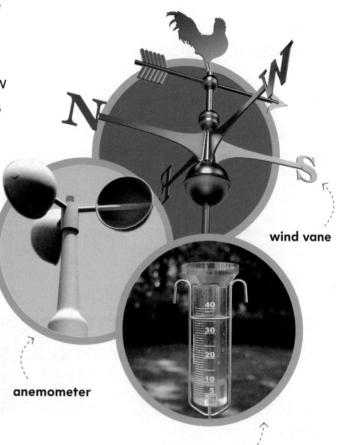

wind vane

anemometer

rain gauge

1. What weather tool would you like to make? (Check one.)

☐ Wind vane   ☐ Anemometer   ☐ Rain gauge

2. Draw your weather tool below. Label your drawing. Write what you would use to make it and how it works. Use the materials you have gathered to make the tool. Then try it out.

Seasons

The Changing Seasons

**Read about the four seasons.
Then try the science investigations.**

Spring! The sun stays out a long time. The air gets warmer. What a nice change from the cold, dark winter. Plants start to bloom. Insects come out. So do animals that count on them for food.

In the **summer**, the weather gets hotter. There's still plenty of food. But little by little, the sun rises later and sets earlier.

In **autumn** the air gets cooler. Leaves change colors. They fall off trees. Some animals gather as much food as they can. Others travel to warmer places.

The days are short and cold. **Winter** has come! Most plants and trees are bare. There's hardly any food to eat. Animals take shelter to keep warm. Others **hibernate**, or go into deep sleep. Maybe they're dreaming of spring!

Investigation 1

Is it spring, summer, fall, or winter? Your calendar may tell you. But you can also find your own clues to the season.

1. Write the current season on the next page. (If you need to, ask an adult for help.)

2. **Think:** What would you expect the weather to be like during this season? What would plants look like? What would animals be doing? What would people be wearing or doing? Write or draw your answers on your recording sheet.

3. Now go outside. Look for clues about each item on your list. Write or draw what you found on your recording sheet.

4. Pick one other season. What do you think would be different if you looked for clues in that season?

Current season: _____

	What I expect	Clues I found
Weather		
Plants		
Animals		
People		

Pick a different season. What do you think would be different if you looked for clues in that season? Write your answer on a separate sheet of paper.

Investigation 2

Keep track of how long or short the days are.

1. Look at the Season Tracker below. How many seasons do you see? How many months?

2. Find the current month. What season is it? (**Hint:** Look to the right of the month.) Are the days getting longer or shorter? During which seasons are the days getting longer? During which seasons are the days getting shorter?

3. The length of a day is the time between sunrise and sunset. Look up the times of today's sunrise and sunset. Write the times in the chart on the next page.

4. **Predict:** In one week, what will be the times for sunrise and sunset? Record your guesses.

5. Wait one week. Find the actual times for sunrise and sunset. Record your results.

6. Do Steps 4 and 5 again for the following week. Look at your chart. What do you notice?

Materials

★ pencil

★ recording sheet (next page)

Season Tracker

January	**Winter** Days are getting longer.
February	
March	**Spring Equinox** Day and night are equal
April	**Spring** Days are getting longer.
May	
June	**Summer Solstice** Longest day and shortest night
July	**Summer** Days are getting shorter.
August	
September	**Fall Equinox** Day and night are equal
October	**Fall** Days are getting shorter.
November	
December	**Winter Solstice** Shortest day and longest night

46

1. What season is it? _____

2. Are the days getting longer or shorter? _____

3. During which seasons are the days getting longer?

_____

4. During which seasons are the days getting shorter?

_____

5. Record the times for sunrise and sunset in the chart below.

	Sunrise time	Sunset time
Today's date: _____		
Date (one week later): _____	I predict: _____ Actual time: _____	I predict: _____ Actual time: _____
Date (two weeks later): _____	I predict: _____ Actual time: _____	I predict: _____ Actual time: _____

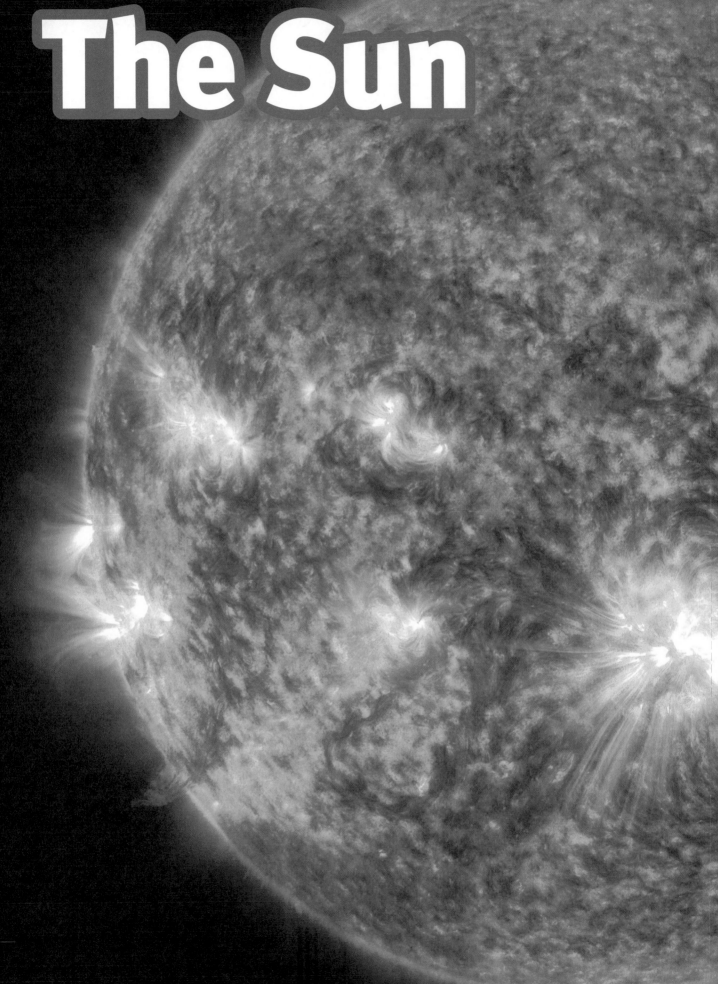

The Sun

Hello, Sun

Read about the sun. Then try the science investigations.

Did you know our sun is actually a star? It is the closest star to Earth. No wonder it looks so much bigger and brighter than other stars in the sky! Like other stars, our sun is a giant superhot ball of gas. It gives us heat and light.

Can you imagine what Earth would be like if there were no sun? It would be cold and dark! No living thing could survive on this planet! We all need the sun to stay alive. Thank you, sun!

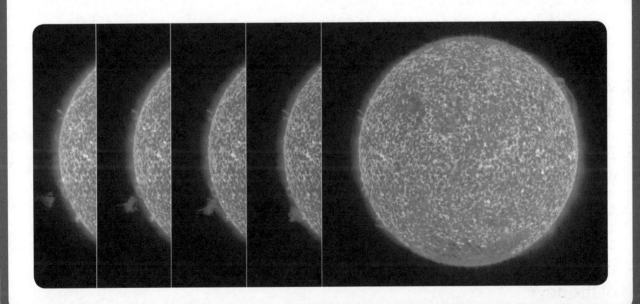

Investigation 1

Track the sun's movement by looking at shadows!

1. Gather the materials you will need.

2. Make a Sun Tracker. Follow the directions on your recording sheet.

3. Take your Sun Tracker outside on a sunny day. Place it on a flat surface. Use a compass to find north. (An adult can help you.) Turn your paper so the word **North** points north. Tape your tracker down.

4. Find the shadow cast by your triangle. Use a dark marker to trace around the shadow. Write the time next to it.

5. Where do you think the shadow will be in 30 minutes? Use a pencil to draw your guess.

6. In 30 minutes, find the shadow again. Trace it with the dark marker. Write the time next to it.

7. Do Steps 5 and 6 again. Which of your predictions was more correct? What do you think will happen to the shadow over the rest of the day?

Materials

★ sheet of tracing paper

★ pencil

★ scissors

★ masking tape

★ sunny day

★ compass

★ dark marker

★ recording sheet (next page)

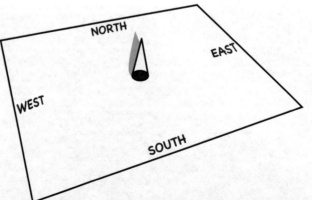

1. Make a Sun Tracker:

- Trace the square and rectangle below onto a separate sheet of paper. Copy all markings.
- Cut out your Sun Tracker (the square).
- Cut out the rectangle. Fold on the solid lines. Overlap the two ends and tape them together.
- Tape the triangle to the middle of the Sun Tracker.

2. Do Steps 3–7 of the investigation. What do you think will happen to the shadow over the rest of the day? Record on a separate sheet of paper.

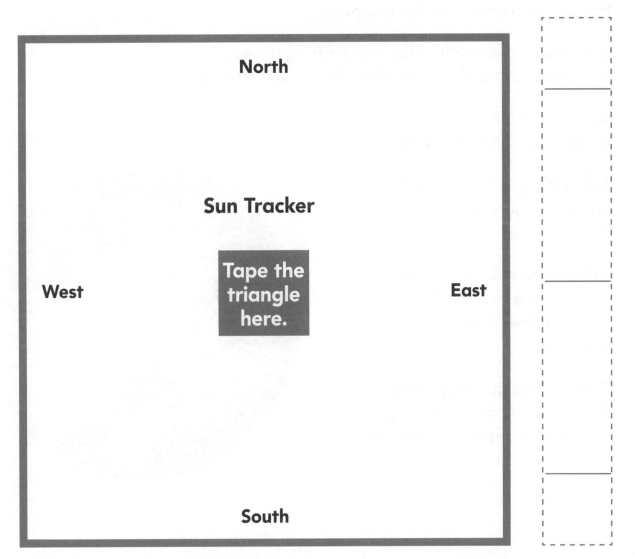

North

Sun Tracker

West

Tape the triangle here.

East

South

Investigation 2

Where does the sun go at night? Find out!

Materials

★ a partner

★ flashlight

★ recording sheet
(next page)

1. Gather the materials you will need.

2. Have a partner hold the flashlight and turn it on. Your partner is the "sun."

3. Stand two steps away from the "sun." You are the "Earth." Your chest is the side of the Earth you live on. Have your partner point the flashlight at your body. Turn so that your left arm is toward the sun. How is this like sunrise? Record your answer on the next page.

4. Slowly turn to your left until you are facing the sun. Which part of the day is this like? Why do you think so?

5. Turn to your left again so your right arm is toward the sun. Which part of the day is this like?

6. Turn left again so your back is now toward the sun. How is this like nighttime?

7. Switch roles with your partner. Now you are the "sun," and your partner is the "Earth." Repeat Steps 2–6.

8. Does the sun go away at night? Why do we have day and night? Record your answers on the next page.

1. Do Steps 2 and 3 of the investigation. How is Step 3 like sunrise?

_____

_____

2. Now do Step 4 of the investigation. Which part of the day is this like? Why do you think so?

_____

_____

3. Do Step 5 of the investigation. Which part of the day is this like?

_____

_____

4. Do Step 6 of the investigation. How is this like nighttime?

_____

_____

5. Does the sun go away at night? Why do we have day and night?

_____

_____

The Moon

Moon Shapes

**Read about why the moon seems to change its shape.
Then try the science investigations.**

The moon is our closest neighbor in space. We often see it in
the night sky. Sometimes the moon looks big and round. Other
times it looks like the tip of your fingernail. Does the moon
actually change its shape?

The moon is round like a ball. It **orbits** (travels around) Earth.
The sun shines on the moon, just like it shines on Earth. As
the moon moves around Earth, we see a different part of the
moon's sunlit surface. When the moon's lit-up side faces away
from us, we call it the **new moon**. When its whole lit-up side
faces us, we call it the **full moon**.

The different shapes of the moon are called **phases**.
What phase is the moon tonight?

Investigation 1

Why does the moon look like it changes shape?
Find out here.

1. Gather the materials you will need.

2. To make your model moon, insert the craft stick into the Styrofoam ball. The flashlight will serve as your sun.

3. Have a partner turn on your model sun. Dim other lights. Stand three big steps from the model sun. Turn so your back is toward the sun.

4. Your head is the model Earth. Hold the model moon in front of you at arm's length. Lift it high enough so your shadow doesn't cover it.

5. Look at the light on your moon. Compare it to the four moon pictures on the next page. Which picture does it look most like? Write "full moon" in the box under that picture.

6. Turn left until the sun is directly on your left. Look at the light and shadow on your moon. Which picture does it most look like? (Notice which side is lit up.) Write "last quarter" in its box.

7. Turn left until you're facing the sun. Which picture does your moon look most like? Write "new moon" in its box.

8. Turn left until the sun is directly on your right. Which picture does your moon look most like? Write "first quarter" in its box.

9. Turn left until the sun is directly behind you again. Your model moon just made one full trip around your model Earth. The real moon takes almost a month to orbit Earth.

Materials

★ craft stick

★ Styrofoam ball

★ flashlight

★ a partner

★ recording sheet (next page)

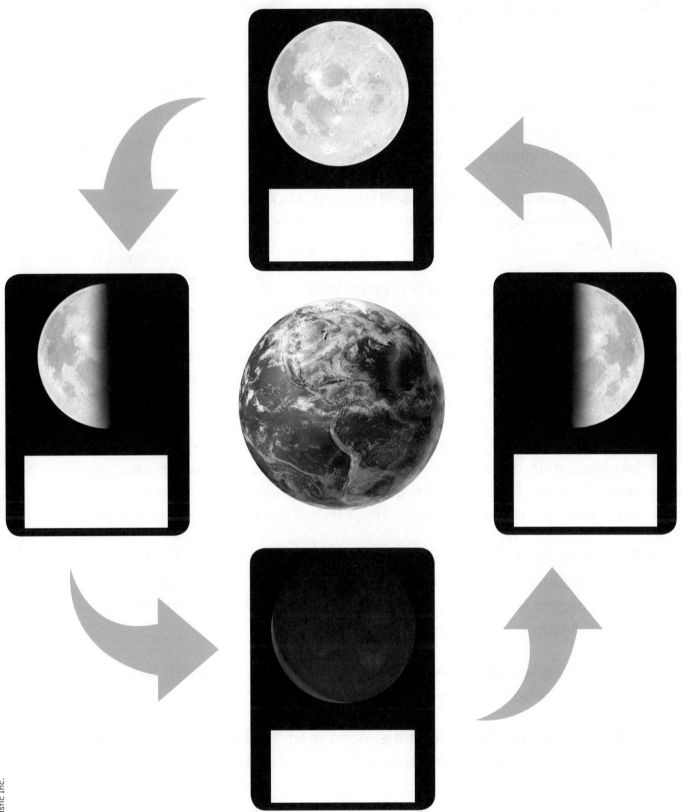

Investigation 2

You have learned about the four moon phases:
Full Moon, Last Quarter, New Moon, and First
Quarter. Now learn about the other phases too!

1. Gather the materials you will need.

2. To make your model moon, insert the craft stick
   into the Styrofoam ball. The flashlight will serve as
   your sun.

3. Use your finger to trace the arrows on your
   recording sheet. The arrows show the order of the moon phases.

4. Look at the four pictures at the bottom of your recording sheet.
   Each one shows another moon phase. These phases happen
   between the first four phases you observed.

5. Ask a partner to turn on your model sun. Dim other lights. Stand three
   big steps from the model sun. Turn so your back is toward the sun.
   Hold the model moon in front of you at arm's length. Lift it high
   enough so your shadow doesn't cover it. You are modeling a Full Moon.

6. Slowly turn to your left until your moon is halfway between the Full Moon
   and Last Quarter phases. Look at the light and shadow on your moon.
   Which picture is it most like? Draw that picture in the box between
   the Full Moon and the Last Quarter.

7. Keep turning left and observing your model. Figure out which picture
   you should draw in each of the other three boxes.

8. **Think:** Why does the shape of the moon seem to change as it orbits the
   Earth? Write your ideas on a separate sheet of paper.

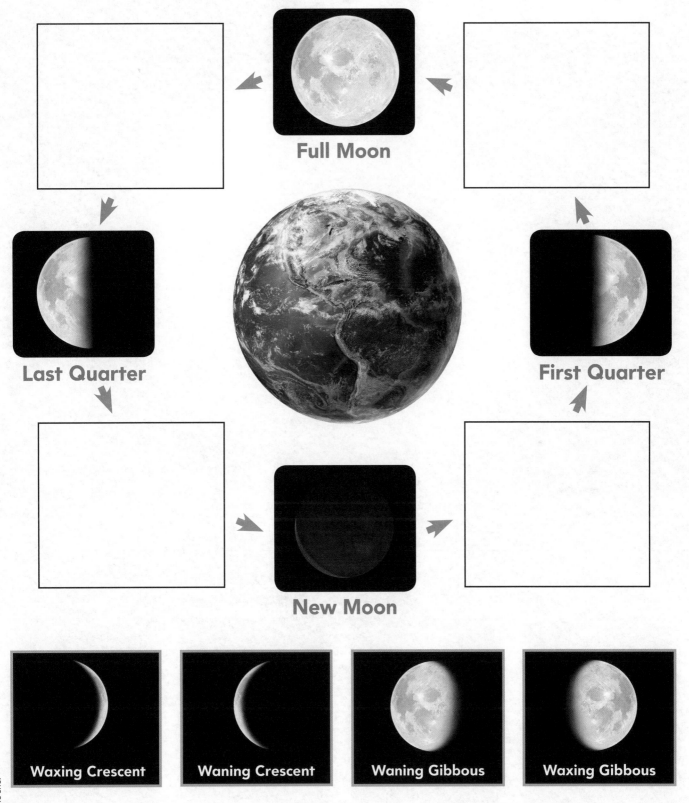

Full Moon

Last Quarter

First Quarter

New Moon

Waxing Crescent

Waning Crescent

Waning Gibbous

Waxing Gibbous

Caves

A person standing in this giant cave!

Read about how a cave forms. Then try the science investigations.

The cave shown at left is one of the biggest caves in the world. How do caves like it form? Water seeps through cracks in the rocky ground. The water slowly **dissolves**, or melts away, some of the rock. Then the water carries the dissolved rock away. This process is called **erosion**. Over thousands of years, the water carves out a larger and larger space. A cave is formed!

This process can also make new shapes inside a cave. Some shapes look like rock icicles hanging down from the ceiling. These are called **stalactites**. Some look like rock piles that grow from the ground. These are called **stalagmites**.

Earth's caves will never be finished. Water will keep shaping them . . . very, very slowly.

Investigation 1

Some rocks dissolve more quickly than others.
Find out how that helps form caves.

1. Gather the materials you need.

2. Squish the clay flat into a strip. Wrap the strip around the sugar cube, as shown below. Make sure there are two openings in the clay—one at the top and one at the bottom.

3. Put the clay and sugar cube against the inside of a bowl, as shown below. The clay and sugar are like two different rocks. Some rocks dissolve more easily than others.

4. **Predict:** What do you think will happen if you drip 100 drops of water onto your rocks? Write your prediction on the next page.

5. Use your dropper to pick up water. Drip it onto your model rocks. Count the drops. After every 20 drops, stop and record what you notice on your recording sheet.

6. How is your model like a cave? How is it different?

Materials

★ clay (about the size of a large marble)

★ sugar cube

★ bowl

★ eyedropper or pipette

★ cup of water

★ recording sheet (next page)

clay

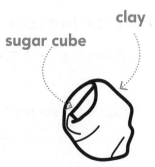

sugar cube   clay

1. Do Steps 2–4 of the investigation. **Predict:** What do you think will happen if you drip 100 drops of water onto your "rocks"?

_____

_____

2. Now do Step 5 of the investigation. Record what you notice.

What I noticed about my rocks	
after 20 drops	
after 40 drops	
after 60 drops	
after 80 drops	
after 100 drops	

3. How is your model like a cave? How is it different?

_____

_____

Investigation 2

Erosion helps form caves. But it can also wash away soil. How can erosion be stopped?

1. Gather the materials you will need.

2. Put the same amount of soil into two milk carton halves. For each one, fill the closed end a little more than halfway full. Let the soil slope down to the open end. Gently pat the soil smooth.

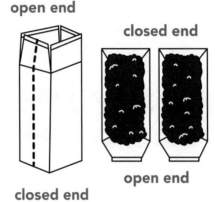

open end

closed end

open end

closed end

3. Set one carton on the ground outdoors. Prop up the closed end on a building block.

4. Put 2 cups of water into a watering can. Pour the water over the soil at the closed end of your carton. Let it drain out completely. What do you notice? Do you see signs of erosion? Record what you see on the next page.

5. **Think:** How can you keep water from eroding the soil? (**Hint:** Plant roots spread through soil and help keep it in place.) Write three ideas on your recording sheet.

6. Choose one of your ideas from Step 5. Gather your inventor's materials and your second milk carton half. Add your invention to stop erosion to the soil.

7. Do Steps 3 and 4 again with your second carton.

8. Compare the two cartons. How well did your invention stop erosion? How could you improve your invention?

Materials

★ soil

★ trowel or large spoon

★ quart-sized milk carton, fully opened on top and cut in half lengthwise

★ building block

★ measuring cup

★ water

★ watering can with sprinkler head

★ inventor's materials: toothpicks, pipe cleaners, index cards, paper towels, plastic forks

★ recording sheet (next page)

1. Do Steps 2–4 of the investigation. Record what you see.

_____

_____

2. How can you keep the water from eroding the soil? Write three ideas.

_____

_____

3. Now do Steps 6 and 7 of the investigation. What do you notice?

_____

_____

4. Compare the two cartons. How well did your invention stop erosion?

_____

_____

5. How could you improve your invention?

_____

_____

Light and Shadow

Tricks of the Light

Read about two animals that use light and shadow to stay alive. Then try the science investigations.

Black Heron

This bird wades in water, looking for tasty fish. But the water's surface acts like a mirror. Sunlight reflects off the surface and into the bird's eyes. The bird can't see the fish below.

But this heron has a trick. It spreads its wings into an umbrella shape. That blocks the light. It makes a dark shadow on the water's surface. The shadow helps it look into the water. When a fish swims into its shadow, the heron can see it and . . . GULP!

Atolla Jellyfish

This jellyfish lives so deep in the ocean, no sunlight reaches it. Creatures there live mostly in the dark. But when they need it, many can make their own light.

The Atolla uses light for protection. If a predator tries to eat it, the jellyfish flashes a ring of blue lights. The lights act like a burglar alarm. Instead of a police officer, the lights attract a large squid. The squid rushes to the rescue and eats the predator. The Atolla is saved!

Investigation 1

Play with light and shadows to make a fun puppet show!

1. Gather the materials you will need.

2. Make a shadow puppet: Draw a person, monster, or animal on the cardboard. Draw a handle from the bottom of the puppet to the bottom of the cardboard. Cut out your puppet.

3. Turn off the lights in the room. Turn on the flashlight. Hold the puppet between the flashlight and a blank wall. Does it make a shadow on the wall?

4. Experiment with your puppet and your flashlight. By moving them around, how can you do each of these "special effects"?

   • Make the puppet's shadow grow bigger.

   • Make the puppet's shadow shrink.

   • Make the puppet's shadow a thin sliver.

   • Make the puppet's shadow move without moving the puppet.

   Record what you did on the next page.

5. Use your shadow puppet and special effects to perform a short show.

Materials

★ pencil

★ cardboard

★ scissors

★ flashlight

★ recording sheet (next page)

Experiment with your puppet and your flashlight by moving them around. Record your observations below.

How did you make the puppet's shadow grow bigger?

How did you make the puppet's shadow shrink?

How did you make the puppet's shadow a thin sliver?

How did you make the puppet's shadow move without moving the puppet?

Investigation 2

What material blocks light the best?
What lets light through? Try this!

1. Gather the materials you will need.

2. Fold an index card in half and cut out a simple shape. Unfold. Do the same with the two other index cards.

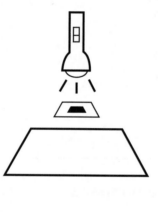

3. Hold one card over a blank sheet of paper. Shine a flashlight through the hole in the card. What do you see on the paper? How much light comes through the hole?

4. Cut a piece of waxed paper a little bigger than the hole. Tape it over the hole in one card.

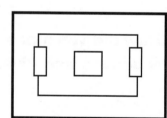

5. Repeat Step 4 with aluminum foil and plastic wrap.

6. Look at your cards. How much light do you think could go through each material? Record your prediction on the next page.

7. Repeat Step 3 with each of your materials. Record your observations on your recording sheet.

8. Imagine your materials were thicker. Which would make the best window? Which would make the best curtain? Which would make the best wall?

Materials

★ 3 index cards

★ scissors

★ blank paper

★ flashlight

★ waxed paper

★ tape

★ aluminum foil

★ plastic wrap

★ recording sheet (next page)

1. Do Steps 2 and 3 of the investigation. What do you see on the paper? How much light comes through the hole?

Check one: ☐ a lot   ☐ some   ☐ none

2. Now do Steps 4–7 of the investigation. Record your predictions and observations in the chart below.

Material	I predict the material will let through this much light	I observed the material let through this much light
Waxed paper	☐ a lot ☐ some ☐ none	☐ a lot ☐ some ☐ none
Aluminum foil	☐ a lot ☐ some ☐ none	☐ a lot ☐ some ☐ none
Plastic	☐ a lot ☐ some ☐ none	☐ a lot ☐ some ☐ none

3. Imagine your materials were thicker. Which would make the best window? The best curtain? The best wall?

_____

_____

_____

_____

Sound

Music Makers and Shakers

Read about how vibrations make music. Then try the science investigations.

When something **vibrates**, it moves back and forth very quickly. Try this: Put your fingers on your throat and say, "vibrations." Can you feel your throat vibrate when you talk? Everything that makes a sound is vibrating, even if you can't see it moving.

How do musicians make their instruments vibrate? A drum is one of the simplest instruments. When you tap the skin of a drum, it vibrates. A harp is played by plucking metal strings. Its strings are different lengths. The short strings vibrate quickly. They make high, squeaky sounds. The long strings vibrate slowly. They make low, deep sounds.

When an instrument vibrates, the air around it vibrates too. The vibrations travel through the air like waves in water. If the sound is loud enough to reach your ears, it vibrates your eardrums. Your ears send signals to your brain. You hear music!

ear

Air vibrates as it leaves the trumpet. The vibrating air hits our eardrum. That's how we hear sound.

Investigation 1

How can you see what sound looks like? Try this!

1. Gather the materials you will need.

2. Cut one side of a balloon from neck to bottom, as shown. Stretch the balloon tight over the top of a can. Have a partner use a rubber band to hold the balloon in place.

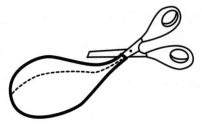

3. Sprinkle a little puffed rice cereal on top of the balloon.

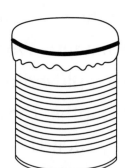

4. Roll the paper or file folder into a megaphone shape, as shown. Make sure there is an opening to speak into. Tape in place.

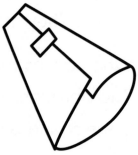

5. Point your megaphone at the balloon and shout. (Don't blow!) Record what happens on the next page.

6. **Think:** What made the cereal do what it did? How do you know? (Use the word *vibrate* in your answer.)

7. Try a shout that's high and squeaky. Then try one that's low and deep. Make loud sounds. Make soft sounds. Do different sounds make the cereal move differently? Record what you notice.

Materials
* scissors
* balloon
* large empty tin can
* rubber band
* a partner
* puffed rice cereal
* stiff paper or file folder
* tape
* recording sheet (next page)

74

1. Do Steps 2–5 of the investigation. Record what happened.

_____

_____

2. Think: What made the cereal do what it did? How do you know? (Use the word _vibrate_ in your answer.)

_____

_____

_____

3. Now do Step 7 of the investigation. Record what you notice below.

Type of sound	What happens to the cereal?
High and squeaky	
Low and deep	
Loud	
Soft	

Investigation 2

Play with sound, then make a musical instrument!

1. Gather the materials you will need.

2. Stretch a rubber band between your thumb and a finger. Pluck it. What do you observe?

3. Stretch the band around a cup, as shown. Pluck it. Does the cup change the sound's volume—how loud or quiet it is?

4. Pluck the rubber band again. Then touch the middle of the rubber band with one finger. This splits the rubber band in half. Pluck one of the halves. Compare the two sounds. Which has a higher sound—the longer (whole) rubber band or the shorter (half) rubber band? Can you play different notes by changing where you touch the rubber band?

5. Stretch the rubber band so it is tighter across the top of the cup. Does this change the sound? How?

6. Use what you have learned to make an instrument. Look at your inventor's materials. **Think:** How will you make your instrument loud enough to hear? How will you make different notes?

7. How does your instrument work? Explain on the next page.

Materials

★ rubber band

★ plastic cup or other container

★ inventor's materials: cardboard boxes, more rubber bands, cardboard tubes, yogurt containers, tape, scissors, string

★ recording sheet (next page)

1. Do Step 2 of the investigation. What did you observe?

_____

_____

_____

2. Now do Step 3 of the investigation. Is the sound louder or quieter?

_____

3. Do Step 4 of the investigation. Which makes a higher sound—the longer (whole) rubber band or the shorter (half) rubber band?

_____

4. Do Step 5 of the investigation. Does stretching the rubber band tight across the cup change the sound? How?

_____

_____

_____

5. Make an instrument. Explain how it works on a separate sheet of paper.

Ice

The Ice Hotel

Read about a hotel made of ice. Then try the science investigations.

There is a hotel in Sweden made out of ice and snow. It is called the Ice Hotel.

In winter, it gets very cold in Sweden. The Torne River turns to ice. People take the ice from the river and mix it with snow. They call this mixture "snice." They make large blocks out of snice. Then they use the blocks to build the hotel. Even the beds are made of snice. The people who stay there sleep in special, warm sleeping bags.

Cool Fact:

How long does it take you to make one snowball? How about 700 million snowballs? That's how much snow is used to build the Ice Hotel!

Investigation 1

When ice is a problem, what's the fastest way to melt it? Find out here!

1. Gather the materials you will need.

2. Put an ice cube in each bag.

3. How can you melt an ice cube fast? Try these:

   • Leave one on a table. (This is your "control." It shows how fast the ice will melt by itself.)

   • Put salt on one ice cube.

   • Crush one ice cube.

   • Think of two other ways.

4. Use the marker to label the bags. Then close them.

5. **Predict:** Which way will melt the ice fastest? Slowest? Record your predictions on the chart on the next page.

6. Wait a few minutes. Check your ice cubes and record what happened.

7. Imagine you are in charge of keeping roads safe. If the roads were covered in slippery ice, what could you do to melt the ice?

Materials

★ 5 ice cubes, same size

★ 5 zip-top sandwich bags

★ ways to melt the ice: 1 tsp salt, something to crush ice with

★ permanent marker

★ recording sheet (next page)

Control

Salt

Crushed

1. Do Steps 2–4 of the investigation. **Predict:** Which way will make the ice melt fastest? Slowest? Number these from 1 to 5 (1 for the fastest, 5 for the slowest) on the chart below.

2. Wait a few minutes. Check your ice cubes. Which ice melted fastest? Slowest? Record in the chart. Number these from 1 to 5 (1 for the fastest, 5 for the slowest).

Ice Melters	My prediction	What happened
Control		
Ice with salt		
Crushed ice		
My 1st idea		
My 2nd idea		

3. If the roads were covered in ice, what could you do to melt the ice? Why? Write your answers on a separate sheet of paper.

Investigation 2

What could you use to keep ice from melting?

1. Gather the materials you will need.

2. **Think:** What material could keep heat away from an ice cube and make the best Ice Keeper? Why?

3. Put one ice cube in each sandwich bag. Press out the air, then zip the bags closed.

4. Fill a cup halfway with crumpled aluminum foil. Push one bag into the foil. Then cover it with more foil. Use the marker to label the cup.

5. Do Step 4 again with crumpled paper and again with the material you chose. Put the fourth bag in a cup by itself. (This is your "control." It shows how fast the ice will melt by itself.) Put all of the cups in a warm place.

6. After 30 minutes, take the ice out of each cup. Record your observations. Then repack the ice in each cup.

7. Look again after 30 minutes. Which Ice Keeper worked best? How can you tell? How could you make an even better Ice Keeper?

Materials

- ★ 4 ice cubes, same size
- ★ 4 zip-top sandwich bags
- ★ 4 plastic cups
- ★ aluminum foil
- ★ marker
- ★ notebook paper
- ★ inventor's materials: cotton balls, paper towels, craft feathers, scraps of fabric, foam packing material, empty chip bags
- ★ warm place
- ★ clock or watch
- ★ recording sheet (next page)

foil

paper

my ice keeper

control

1. What material do you think would make the best Ice Keeper? Why?

_____

_____

2. Do steps 3–7 of the investigation. Record your observations below.

Ice Keepers	After 30 minutes	After 1 hour
Foil		
Crumpled paper		
Your Ice Keeper		
Control		

3. Which Ice Keeper worked best? How can you tell?

_____

_____

4. How could you make an even better Ice Keeper? Write your ideas on a separate sheet of paper.

Wind Power

Powerful Push

Read about wind. Then try the science investigations.

Wind is moving air. Its force can blow your hat off your head. It can turn your umbrella inside out. Very strong storm winds can even blow the roof off a house.

We can't control the wind. But people have learned how to use the wind's power. We can use it to sail boats. We can use it to keep kites high in the air. We can even use it to make electricity for our homes. How else can we use wind?

Investigation 1

Use wind power to give this toy a push!

1. Gather the materials you will need.

2. Look at the square below. Use it as a guide to make your own Spin Wheel. Follow these steps:

   • Use a ruler to draw a square that measures 4 inches tall by 4 inches wide on a separate sheet of paper.

   • Copy the shapes and dashed lines in this picture onto your own square.

   • Cut out the square.

3. With the printed side up, fold it in half so the bottom edge meets the top edge. Unfold.

4. Fold it in half again so the right edge meets the left edge. Unfold.

5. Flip the paper over so the blank side is up. Fold one corner to its opposite corner. Unfold.

6. Fold so that the other two corners meet. Unfold.

> **Materials**
> ★ ruler
> ★ pencil
> ★ sheet of paper
> ★ scissors
> ★ recording sheet (next page)

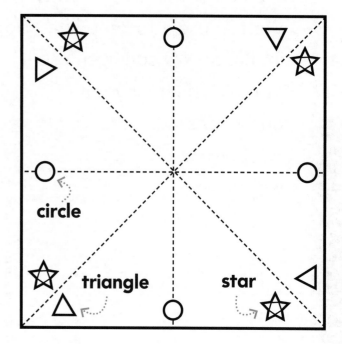

7. Flip the paper over. Gently push on all four circles. The middle of the paper should poke up in a point.

8. Hold a pencil with the tip pointing up. Balance the folded paper on the tip.

9. You can blow air at the Spin Wheel to give it a push. **Predict:** Which parts of the Spin Wheel could you push (or blow) to make it spin?

1. Do Steps 2–7 of the investigation.

  Predict: Which parts of the Spin Wheel could you push (blow on) to make it spin? Check any of the boxes or none.

☐ the stars ☐ straight down on the top

☐ the circles ☐ straight up from the bottom

☐ the triangles

2. Test your predictions. Which worked best to turn the Spin Wheel?

_____

_____

_____

_____

3. Try to make the Spin Wheel turn slowly. Try to make it turn quickly. How did you do it? Use the words *push* and *power* in your writing.

_____

_____

_____

_____

Investigation 2

What would help your toy catch the wind better? Try this!

1. Gather the materials you will need.

2. Follow these steps to make a Super Spinner Cone:

   • Trace the circle below onto a separate sheet of paper. Copy all lines and shading.

   • Cut out the circle. Then cut along the dotted lines.

   • Line up the cut edge with the solid line. The gray pattern will be covered up. The two halves of the star will line up.

   • Tape down the edges.

3. Put your cone over a pencil point. Blow on it gently. What happens?

4. **Think:** How could you change your cone to make it spin like a Spin Wheel? (**Hint:** You can cut and fold it. You can tape new pieces on to it.) Write down three ideas that might work.

5. Pick one of your ideas and test it. (**Hint:** It's important to keep your changes balanced. If you add things to only one side, the cone will fall off the pencil!) If your idea works, think of ways to make it even better. If it doesn't work, pick a different idea to test.

6. Write about your best design on a separate sheet of paper. Why can it spin when you blow on it? Did you run into any problems? How did you fix them?

Materials

★ pencil

★ sheet of paper

★ ruler

★ scissors

★ tape

★ inventor's materials: paper, index cards, straws, pipe cleaners, egg cartons, plastic wrap, store-bought feathers

★ recording sheet (next page)

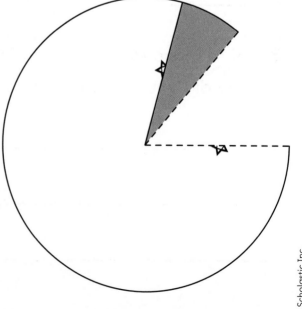

1. Do Steps 2 and 3 of the investigation. What happens when you blow on your Super Spinner Cone?

_____

_____

_____

2. Think: How could you change your cone to make it spin like a Spin Wheel? Write down three ideas that might work.

Idea 1: _____

_____

_____

Idea 2: _____

_____

_____

Idea 3: _____

_____

_____

Forces

Let's Roll

Read about a race car that has no motors. Then try the science investigations.

Every summer in Akron, Ohio, children race in the All-American Soap Box Derby. They race in cars they built themselves.

Every racer gets a kit from the Soap Box Derby. The kit comes with the car's body, brakes (for stopping), a steering wheel (to turn the car), and a helmet. Wheels come separately.

But the cars don't have motors. So what makes them move? **Gravity**—the force that pulls down on us—pulls the cars down the hill!

Racers have to figure out how to make their cars go fast. They can ask an adult for help. What do you think can help these cars go faster?

Did you know?

Force is a push or a pull that makes something move. **Gravity** is a force.

Investigation 1

Find out if an empty bottle has more or less push than a full bottle!

1. Gather the materials you will need.

2. Make a ramp using a folded game board and blocks. Tape the pieces in place.

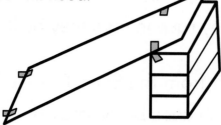

3. Hold your hand at the bottom of the ramp. Have a partner roll the empty bottle down the ramp. Let the bottle push your hand. How did the push feel?

4. Repeat Step 3 with the full bottle. Which bottle gave your hand a bigger push? Why?

5. Place the cereal box in front of the ramp, as shown. **Predict:** What will happen if you roll the empty bottle down the ramp? Test your prediction. Record your results.

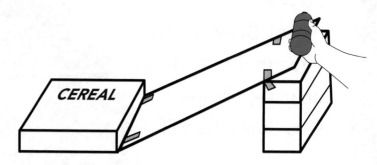

6. Put the cereal box back in front of the ramp. **Predict:** What will happen if you roll the full bottle down the ramp? Test your prediction. Record your results. How much did the box move with the bigger push? Why do you think this happened?

Materials
★ game board
★ blocks
★ masking tape
★ a partner
★ 2 plastic bottles, one full and one empty
★ empty cereal box
★ recording sheet (next page)

1. Do Steps 2 and 3 of the investigation. How did the push feel?

_____

2. Now do Step 4 of the investigation. Which bottle gave your hand a bigger push? Why do you think that is?

_____

3. Do Step 5 of the investigation. **Predict:** What will happen if you roll the empty bottle down the ramp?

_____

4. What happened when you rolled the empty bottle?

_____

5. Do Step 6. **Predict:** What will happen if you roll the full bottle down the ramp?

_____

6. How much did the box move with this bigger push? Check one:

☐ farther   ☐ the same   ☐ not as far

Why do you think this happened?
Write your answer on a separate sheet of paper.

Investigation 2

How can you make a bigger push? Try this!

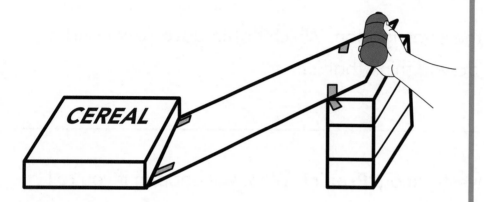

CEREAL

1. Gather the materials you will need.

2. Make a ramp using a folded game board and blocks. Tape the pieces in place. Put the cereal box in front of the ramp. Roll the full plastic bottle down the ramp. If the box moves, put a piece of tape on the floor to mark its new position.

3. **Think:** What one thing could you change to make the box move farther? Think about the box, the bottle, and the ramp. What could be changed about each one? (For instance, in Investigation 1 you used a light, empty bottle, then changed to a heavy, full bottle.) Pick one idea to try.

4. Now make up an experiment to test the change. It's important to test only one change at a time. Everything else in your experiment should stay the same. What one thing will you change? What will you keep the same?

5. Write the steps of your experiment on your recording sheet. Then do your experiment! Record what happened.

1. Do Steps 2 and 3 of the investigation. What could you change to make the box move farther? List your ideas.

_____

_____

2. Pick one idea. I will try this change:

_____

3. Write down the steps of your experiment.

_____

_____

_____

_____

4. Do your experiment. Record what happened.

_____

_____

_____

_____

Hands-On Hints

Background information plus tips and expected results for each science investigation

Life Science

Earth Science

Physical Science

Animal Babies

Background

The animal kingdom has all kinds of parents. Some animals, such as penguins and monkeys, take care of their young until they are old enough to survive on their own. Other animals, such as newts, lay their eggs and leave them behind. When the young hatch, they have to fend for themselves.

Hands-On Hints

Investigation 1

Introduce this activity by talking about the idea of a *model*—something people make to represent bigger or more complicated things. Give examples of models (e.g., a model train set, a building made of blocks, a computer game that simulates a car race). Explain that making a model can be a good way to learn about the real thing. With this activity, your child is going to model a penguin's behavior. In a way, your child will become a model penguin in order to learn about real penguins.

If you have sand available, you can substitute that for rice in the activity. Stiff or slick shoes may make it harder to keep the "egg" in place. Your child may want to either take off his or her shoes or put socks over slick shoes (to increase the "gripping" friction between egg and feet).

Slowing down and taking small, "waddling" steps may help keep the egg on feet. So may repositioning feet—tipping up toes, for example. Penguin fathers carry their eggs on their feet to keep them warm (by keeping them off the ice and next to their warm bodies).

After the activity, ask your child: *How are penguin dads like human dads?* (They help care for their young.) *How are they different?* (Accept reasonable answers.) *Do human fathers ever carry their children on their feet?* (Sometimes in play) *How do human fathers carry children?* (In their arms, on their backs, in baby carriers) *Why don't penguins carry their babies like humans do?* (Accept reasonable answers. For one thing, they have wings, but no hands.) Point out that penguin legs are very short. That means their feet are near their tummies, so penguin fathers can lap their tummies over an egg to help keep it on their feet. Ask: *Why can't human fathers do that?* (Their legs are too long.)

Investigation 2

In this activity, your child will design a baby bird that stands out in a habitat and another baby bird that blends in.

Explain that many animal babies, such as seals, turtles, and deer, are camouflaged to hide from predators. Animal babies are not the only ones that rely on camouflage. Many adult animals do as well. Penguins, for instance, may stand out in the snow. But in the water, where many of their predators live, they are actually well-camouflaged. Seen from above, penguins' black backs and wings blend in with the dark water below. Seen from underneath, their white bellies help them blend in better with the sunlight from above.

Plants

Background

Scientists estimate that there are about 300,000 to 400,000 different kinds of plants on our planet. They can be as small as duckweeds, which are about a tenth of an inch long, or as large as giant sequoias, which can grow to more than 300 feet high.

Plants play an important role in our ecosystem because they are **primary producers**. In other words, they can make their own food through photosynthesis. Almost all of the food chains on our planet start with plants. Of course, humans rely on plants for other things as well, such as wood, clothing, and even medicines.

Hands-On Hints

Investigation 1

The paper towels should be moist, but not dripping. One layer of wax paper rolled around Leaf 3 is sufficient.

After 24 hours, our flat paper towel (Leaf 1) was dry and the rolled towel (Leaf 2) was damp inside, but dry on the outside. But the towel wrapped in wax paper (Leaf 3) was still damp. Like some cactus leaves, Leaf 3 kept water loss to a minimum because of its reduced surface area and waxy covering.

After your child has observed how the towels change, you may want to introduce the word *evaporation*—the process in which a liquid becomes a gas.

Investigation 2

Depending on how your child wants his or her seed to travel, set up one of the three Seed-Testing Stations below to test the seed.

- Glide Test Station – You'll need an open space. Consider using an electric fan for a consistent wind source.

- Float Test Station – Put a dishpan of water on a desk. Place a towel or newspaper under the dishpan.

- Stick Test Station – Get a stuffed animal or use fuzzy socks for the animal.

Provide a variety of materials for your child to use on the seed. Good bets include: cotton balls, pillow stuffing, craft feathers, plastic wrap, balloons, foam packing peanuts, aluminum foil, toothpicks, bits of Velcro. Don't forget tape and glue.

Why do seeds need to travel? Plants need to soak up water and nutrients to survive. But they can't walk around to get them. They have to rely on the soil in which they're growing. And they have to share the water and nutrients with other plants growing around them. What would happen if all of a plant's seeds fell right beneath it? The seeds would all sprout right around the mother plant, and there wouldn't be enough water and nutrients to go around.

Needs of Living Things

Background

Moto is a serval kitten that was found in 2005 after a wildfire swept through the grasslands of the Masai Mara Reserve in Kenya. A serval is a medium-sized wildcat with a tan or yellowish coat with black spots and large ears. It has a long, thick tail and very long legs for its size. Servals live in the marshes, grasslands, and woodlands of sub-Saharan Africa. Their diet includes rodents, small birds, reptiles, frogs, fish, and insects.

Hands-On Hints

Investigation 1

You may need to help press out the air from the bags and seal them. If you need to move the bags, place them on a tray. The less disturbed the bags are, the easier it will be to see the air bubbles forming in the "Food and Water" bag.

If possible, have a slice of leavened bread and unleavened bread in a plastic bag for your child to look at and compare with a hand lens. Can he or she see evidence of yeast "eating" flour during the making of the leavened bread? (The "bubbles" in the bread)

The gas released by the yeast as it processes the food is carbon dioxide. Yeast needs water to function, much like plants and animals need water to function. Yeast needs food, like animals do. It can't make its own food using the energy from sunlight, like plants can.

After your child has finished with the activity, you can either compost the yeasty mixtures or throw away the bags unopened. Don't pour them down a drain.

Investigation 2

Your child may adapt the first yeast activity for the food test or may come up with something new. The important thing is that your child controls variables and has a plan for deciding which food worked better (perhaps noticing the amount of gas released by the yeast while processing the food).

Pure sugar works very well and likely will result in producing the most impressive amount of gas.

Habitats

Background

Australia used to be home to millions of koalas. In 2015, the Australian Koala Foundation estimated that there are less than 80,000 koalas left in the wild.

European settlers who moved into Australia 200 years ago cut down eucalpytus forests and hunted koalas for their soft fur. In some areas, the animals have almost disappeared. Today, hunting koalas is illegal. But they're still losing their habitat to people. Living so close to people also puts koalas in danger. Many are hit by cars or attacked by pet dogs.

Hands-On Hints

Investigation 1

How you do this activity will depend on your community park or backyard, but even a city's paved play space will likely have some signs of life. Look in the cracks or on the edges for small plants and insect life. Look for leaves or seeds that have blown in from outside.

Talk over safety rules with your child before going outside. Details will depend on your park, but should probably include:

- Never put your fingers where you can't see them—like inside a hole in a tree trunk or the ground.

- Do not touch or otherwise disturb any animals, including insects.

- Signs of animal life like molted feathers, spider webs, and animal scat (poop—yes, that's a sign of life!) should be left where they are.

- Don't eat anything you find— for instance, berries.

- Add any necessary warnings about poison ivy, nettles, thorns, and so on.

If you wish, help your child gather some signs of life for closer examination at home. Feathers, half-eaten acorns, and such can be collected in zip-top bags or placed on a tray. You can use clippers, scissors, or a trowel to carry away a plant sample. Put it in water to keep it fresh. Insects can be kept in clear containers with air holes. Include a sample of the plant they were found on and let them go at the end of the day. A hand lens will aid in closer observation.

Investigation 2

Encourage your child to think as big as he or she wants. If your child decides to imagine how the park could accommodate a whale, consider it a good opportunity for exercising problem-solving abilities! Your child will be thinking about animal needs and ways humans can lessen our impact on other living things. A younger child can rely on prior knowledge and imagination. Encourage an older child to research his or her subject.

Depending on your geography, birds and butterflies are both animals whose needs can be met with feeders or plantings. Good resources would include a local chapter of the National Audubon Society or the American Horticultural Society website.

Pollinators

Background

Bees pollinate about one-third of all the world's crops. Without bees, we could very well have a severe food shortage. Since 2006, tens of millions of bees have mysteriously disappeared in the United States. Nearly half of all the bee colonies in the U.S. have vanished.

Scientists have a few theories as to why bees are disappearing. One reason may be pesticides, which farmers use to get rid of harmful insects. But they can kill bees as well. Bees may also be suffering from poor nutrition. To stay healthy, they need to drink nectar from many different kinds of wildflowers. But many of the places where wildflowers once grew have been cleared to make room for farms, buildings, and roads.

What can be done to help bees? The government has created a group called the Pollinator Health Task Force, which plans to plant wildflowers on 7 million acres of land across the country. Children and their families can also help by planting "bee-friendly" flowers at home.

Hands-On Hints

Investigation 1

The more colors of glitter you have, the easier it will be to see how widely the pollen has spread. (Make sure each flower starts out with only one color.)

Investigation 2

Explain to your child why it's helpful that different pollinators have specific preferences when it comes to flowers. Take hummingbirds, for example. They like flowers that have long, red tubes. Say they visit only those favorite flowers. That means they will gather pollen only from long red flowers. And they will leave pollen behind only in long red flowers. That means those red-flowered plants have a better chance of getting pollen from other plants just like them. And that's exactly what they need to make seeds.

Decide with your child how fancy to make the flower. Your child can simply draw and color a flower on white paper. Or you can provide materials such as construction paper, colored tissue paper, pipe cleaners, and so on. With a younger child, you might want to model how to curl a rectangle to make a tube.

If your child is not allergic or sensitive to smells, you may want to invite him or her to add a scent to the flower. To add a sweet, spicy, or fruity smell, you can offer a cotton ball dabbed in liquid like vanilla extract, lemon extract, or scented liquid soap. Your child can swab the flower with the scent or add the cotton ball to his or her design.

Weather

Background

You can demonstrate cloud formation using a clear glass container with a wide opening, a frozen cold pack (or a bag of ice), very warm water (at least as warm as steamy bath water—you may want to test this activity ahead of time to make sure your water is warm enough), and a can of spray deodorant.

Explain to your child that to make a cloud, you need water vapor, cold temperature, and tiny particles for water droplets to form around. Explain that some of your very warm water will quickly evaporate to become water vapor. The cold pack will provide cold temperature. The can of spray deodorant will provide the tiny particles for droplets to form around. Pour about an inch of very warm water into the glass container. Place a hand over the opening for about 10 seconds to allow the water vapor to gather. Give a quick spray of the deodorant into the container and place the frozen cold pack on top. Fairly quickly, you should see a thick cloud forming in the container. If you lift the ice pack, some of your "mini cloud" will start to escape.

Hands-On Hints

Investigation 1

Choose a regular time at the beginning of the day to record cloud data. Have your child record the weather at the same time each afternoon. Remind your child to notice the weather at bedtime also.

Results for this activity will depend on your location, the time of year, and what kind of weather front is moving through your area. You can choose to do this activity on four consecutive days, or you can wait for days that have one of the four common types of clouds listed on the Cloud Tracker chart. You might also watch weather forecasts to scope out days that might have "interesting" weather.

Data may show that certain clouds precede certain kinds of weather. Or your child may find that there was no rain all week, even though he or she saw many kinds of clouds. If the weather data your child collects is contradictory, remind him or her that weather scientists have had to collect years of data from all over the world to find reliable patterns. This activity allows your child to observe and consider weather like a scientist.

Investigation 2

After your child designs a weather tool, invite him or her to make a model. Then have your child try it to see if it works the way he or she thought it would. Encourage your child to make changes, if necessary, and test it again.

Seasons

Background

The Earth is tilted about 23 degrees on its axis. As it revolves around the sun, half of the Earth leans toward the sun while the other half is tilted away. This is what gives us the four seasons. The Northern Hemisphere, where the United States is located, leans toward the sun from June to September. During these summer months, the days are long and the weather is warm. At the same time the Southern Hemisphere, where Australia is located, is tilted away from the sun. The days are short and the weather cold—winter. From December to March, the Southern Hemisphere is tilted toward the sun and thus enjoys summer, while the Northern Hemisphere experiences winter.

Hands-On Hints

Investigation 1

As an extension, you can make a season "clock" with your child. Take a piece of poster board and use two crossed lines to divide it into four quadrants. Cut a large clock hand out of construction paper and use a brass fastener to attach it to the middle of the poster board. Label each of the four quadrants with a season, going in clockwise order. Invite your child to draw and cut out "clues" to each season, then add them to the correct quadrant. This is a good opportunity to emphasize the seasons where you live. Feature local plants and sights. No need to add snowflakes or fall leaves if they don't apply.

Investigation 2

For simplicity's sake, plan this activity for a two-week period when there will be no shift from or to Daylight Saving Time. Avoid the weeks around the winter and summer solstices as well.

The Season Tracker is designed to help your child recognize seasonal patterns, such as the repeating order of the seasons and changes in the length of days. Steps 1 and 2 of this activity help familiarize your child with the Season Tracker. Steps 3–6 span over two weeks and require your child to look up sunrise and sunset times for your location on a given day. These are often available from weather websites and apps, or from a direct search-engine query (for example, "sunrise Ames Iowa").

Once you have found the sunrise and sunset times, your child can use the Season Tracker to find out whether the days are getting longer or shorter and make a prediction about when sunrise and sunset will occur in one week. You may want to have a discussion to cement your child's understanding that a lengthening day would mean the sunrise and sunset would be getting further apart. When you then look up that day's times and add them to the chart, your child can notice two things: First, the sunrise got earlier (or later, depending on the season) and the sunset got later (or earlier). Second, he or she can see how much change in minutes there was. These facts should help your child make a more accurate prediction for the following week.

Sunrise and sunset times—and the changes you will see in them—are determined by four factors:

- your hemisphere (the Season Tracker applies to the Northern Hemisphere, where the United States is)

- the time of the year

- your latitude, or distance from the equator (the farther you are from the equator, the larger the changes in the length of the day over the course of a year)

- your longitude, or distance east or west from the prime meridian (an imaginary line in England). This is the only factor that is human-made. Originally, noon was defined by when the sun was at its highest point in the sky. That was different for every longitude, because it depends on the Earth's rotation. But once the world was divided into time zones, for most people, the official noon no longer matched the time when the sun was at its highest. This affects sunrise and sunset times as well.

If your family has a link to or interest in another place, you may want to track that place's sunrises and sunsets at the same time.

The Sun

Background

The image in the article is the sun as seen by NASA's high-tech instruments. They record wavelengths of light that humans can't see, such as ultraviolet. NASA colorizes the images to make them more appealing and "sun colored."

Hands-On Hints

Investigation 1

Ahead of time, scope out a hard, flat surface that will not be in shadow during this activity. If the only open areas available aren't hard, have your child tape the Sun Tracker to a clipboard or piece of cardboard. Masking tape is recommended because you can write on it. If you don't have a compass, try a compass phone app.

This activity can be adjusted to the flow of your day—for instance, it may be more convenient for you to check shadows every 45 minutes. The wonders of modern weather forecasting can help you choose a day with steady sun. However, if a band of clouds rolls in and prevents you from checking shadows at the appointed time, just check them when it gets sunny again. Discuss how your child's prediction (based on a half hour's sun movement) relates to his or her measurement.

Your child's first "prediction" (Step 5) can be a flat-out guess. He or she has no information other than where the shadow is right then. When your child makes a second prediction (Step 7), he or she will have evidence of the direction and the speed of the shadow's movement. Your child may be able to use this to make a more accurate prediction. Once your child has a third measurement, he or she can be more certain that the shadow will keep moving in that direction for the rest of the day.

Investigation 2

Before starting the activity, ask your child: *Why do we have day and night?* You may want to record his or her response so you can revisit it at the end of the activity and discuss any changes in his or her understanding.

You may wish to dim or turn off your lights for this activity. Your child will need light for reading and recording answers, but dimmer lights will help him or her see the "sunlight" on the body better. Tell your child not to point the flashlight at others' faces.

Guide your child in making connections between the activity in Investigation 1, where the sun seems to move across the sky, and this activity, which demonstrates that the Earth turns to make day and night.

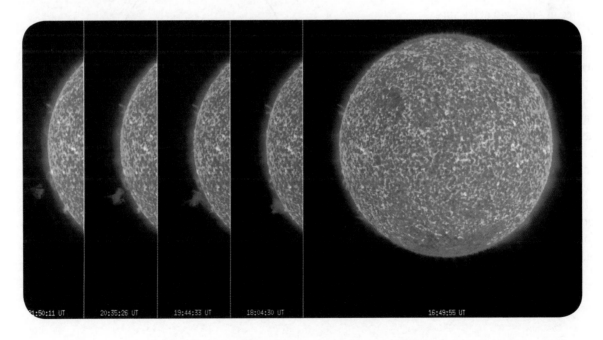

The Moon

Background

The time from the new moon (the dark phase) until the next new moon is about a month! The moon is "tidally locked," which means the same side points toward Earth as it orbits around us.

Hands-On Hints

If necessary, review "left" and "right" with your child before doing these activities.

Model Moon: For each activity, you need a small white ball on a stick. You can buy a Styrofoam ball at a crafts store and stick it on the pointed end of a pencil. Or you can take a Ping-Pong ball and use strips of transparent tape to attach it to the eraser end of a pencil, as shown at right.

Model Sun: This can be a flashlight, a lamp, or even a lit-up computer screen. Your child will be facing this light source at times, so make sure it's not uncomfortably bright. You can tone down a bright flashlight with layers of wax paper. We found that a piece of wax paper folded in half three times (to make eight layers) and attached to an LED flashlight with a rubber band worked well. If you use a flashlight, the beam may be narrow enough that you need to have one partner act as the sun, keeping the light focused on the model moon. Since light sources vary, test yours ahead of time.

Investigation 1

Starting with the full moon at the top of the diagram, the observed phases run counterclockwise: last quarter, new moon, first quarter (Steps 5–9).

Investigation 2

Going counterclockwise from the full moon, the phases should be placed in this order: full moon, waning gibbous, last quarter, waning crescent, new moon, waxing crescent, first quarter, waxing gibbous (Steps 6 and 7).

The moon seems to change shape (Step 8) because we usually see only the lit-up part. As the moon orbits Earth, different amounts of the lighted part are visible from our planet.

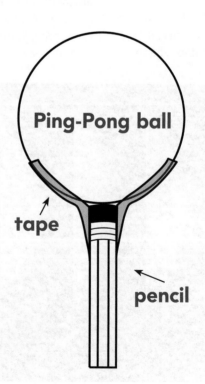

Ping-Pong ball

tape

pencil

Caves

Background

Caves can be formed by several different processes, but the solutional cave formation described in the article is the most common. Rainwater picks up carbon dioxide from the atmosphere and from decaying matter in the soil. Some of the water and carbon dioxide combine to make a weak acid called **carbonic acid**. This acidity allows the water to dissolve certain kinds of rock, including limestone. A cave is generally filled with water during its initial creation. After most of the water drains, dripping water can build formations like stalactites and stalagmites, and flowing water may further shape the cave by erosion.

Hands-On Hints

Investigation 1

In this model, the sugar cube is limestone and the clay is a less-dissoluble rock, like sandstone.

If you don't have an eyedropper or pipette, you can make a straw dropper instead: Fold over one end of half a straw, then fold it again and tape in place, as shown below. Make sure the folded end is airtight.

You may want to have your child practice using the

straw dropper. Have him or her dip the open end into some water and squeeze the sides of the straw to push out some air. Stop squeezing so the straw pulls in some water. Lift the straw and slowly squeeze out drops one at a time.

When wrapping the clay around the sugar cube, have your child leave one edge of the cube sticking up. The opening should be tilted partway up.

The sugar cube will be mostly or completely dissolved by 100 drops of water. Your child may observe that a sharp edge has been rounded, the cube is getting smaller, or a hole has developed in the cube (Step 5). The remaining clay will be shaped like a cave and will have been formed by water. It will be different from a real cave in many

ways, including size and materials, but most importantly in the speed at which it was formed (Step 6).

Investigation 2

Cutting milk cartons requires sharp, strong scissors, so it's an adult's job. The thick carton bottoms may require use of a utility knife. Soil can be bagged potting soil or soil from outside. The important thing is that it is fairly uniform, so that the soil remains a controlled variable.

You can do the first erosion trial (Steps 3 and 4) as a demonstration and have your child compare it to his or her own trial (Step 8). Consider measuring out the soil for each trial to standardize the amount in each carton.

Use a watering can with a sprinkler head so water falls more gently on the soil. You can also poke a dozen small holes in the bottom of a large paper cup and pour water through that.

Signs of erosion include soil that has washed out of the container and channels cut into the soil by running water (Step 4).

Along with plant roots, other real-world erosion-control methods include using sand bags, mulch, and matting/netting.

Light and Shadow

Background

The black heron is a wading bird indigenous to West Africa. Besides helping it see better by cutting down on reflection, the heron's shadow may actually attract fish. The dark area created by the shadow may draw fish by giving them a false sense of safety.

Dr. Edith Widder, a biologist and deep-sea explorer who is a leading researcher in the field of bioluminescence, has found deep-sea creatures that use lights to lure food, scare predators, attract mates, and—as with the Atolla jellyfish—lure larger predators to get themselves out of danger. But she says there is still much to learn about what bioluminescent creatures are doing—and saying—with their flashing lights.

Hands-On Hints

Investigation 1

Depending on the cardboard's thickness, you may want to help your child cut the cardboard. Card stock, which is thinner and easier to cut than regular cardboard, will also work. To make a card-stock puppet less floppy, replace the handle with a taped-on straw or popsicle stick.

Your child can make the puppet's shadow bigger by moving the puppet closer to the flashlight—or the flashlight closer to the puppet. To make the shadow smaller again, put more distance between the puppet and the light. Make the puppet's shadow a thin sliver by giving the puppet a quarter turn, so the light hits the edge of the cardboard. Your child can make the puppet's shadow move by moving the flashlight, even if the puppet stays still. The farther from the wall the puppet is at the time, the more it will move.

Investigation 2

You may wish to share this vocabulary with your child: If light can pass easily through a material (like plastic wrap), we say it is *transparent*. If light cannot pass through a material (like aluminum foil), we say it is *opaque*. If a material lets only some light through (like waxed paper), we call it *translucent*.

As an extension to the activity, you might want to provide an assortment of other materials to test, such as plastic shopping bags, cloth scraps, tissue paper, colored cellophane, and clean food wrappers.

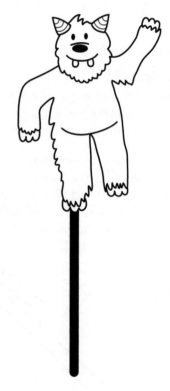

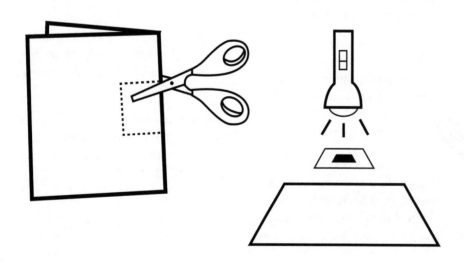

Sound

Background

Sound is a form of energy, just like light, heat, and electricity. Sound is produced when things move back and forth, or **vibrate**, creating sound waves. Sound waves can travel through different kinds of materials, such as metal, water, and air. When sound waves travel through air and reach our ears, we can hear the sound.

Hands-On Hints

Investigation 1

Make sure there aren't jagged or sharp edges where the lid was removed from the can. If there are, cover the edges with tape. Larger balloons are easier to stretch across the cans. The tighter the balloon "drumhead," the more easily it is to see cereal movement.

A loud shout will vibrate the balloon skin, causing the cereal to "jump" (Step 5). High sounds come from faster vibrations, and low sounds come from slower vibrations. Loud sounds come from bigger vibrations, and soft sounds come from smaller vibrations. Your child may see these differences reflected in how the cereal moves (Step 7).

Bonus: Challenge your child to design and test an experiment to make sure that it's the sound making the cereal "jump," not their breath. (For example, find a different way to make noise, such as by playing music through a speaker or clapping two wooden blocks together.)

Investigation 2

Your child will probably see the rubber band vibrating and may hear a quiet sound (Step 2). Putting the rubber band around a cup and plucking it will make a much louder sound (Step 3). Placing a finger in the middle of the rubber band and plucking half of it will make a higher sound (Step 4). Moving one finger around while plucking with the other or stretching the rubber band (Step 5) will both change the pitch.

Depending on materials you have available, your child could make a percussion instrument (using a can or plastic container), a stringed instrument (using rubber bands or string and a box), and more.

Ice

Background

The Ice Hotel is around only from December to April. Every year, the hotel is rebuilt with new architecture and themes. The ice used to build the hotel comes from the Torne River. Builders use 1,000 tons of ice and 30,000 cubic meters of "snice" to fortify the structure. The temperature inside the hotel is a frigid 23°F (−5°C).

Hands-On Hints

Investigation 1

It's important to make the ice cubes the same size. We used 1 tablespoon of water to make each ice cube.

Of the options suggested in the activity, we found that crushing the ice worked best. Why? By increasing the ice's surface area, it comes into contact with more warm air.

When salt is sprinkled on ice, as on an icy road, the small amount of pressure that it exerts causes a little of the ice to melt. Then some of the salt dissolves in the liquid water. This salt solution freezes at a lower temperature than pure water. (Other water-soluble solids also will act to lower the freezing point.) So using salt on icy roads helps melt the ice and keep it from refreezing. (Note: Salting ice can have an impact on soil and plants. While it is necessary for safety, it's important to use only as much as needed.)

Sand seemed to insulate the ice and kept it from melting. But the reddish-brown sand used on icy roads in some areas absorbs heat from the sun, helping to melt ice while providing traction for vehicles.

Of all the ways we tried, adding hot water worked fastest because heat came into direct contact with the ice.

Investigation 2

Once again, make the ice cubes the same size. When putting the ice cubes in the cups, make sure each cube is surrounded by the "Ice Keeper" insulating material. You may want to put a piece of tape over each cup to keep the insulating material firmly around the ice cubes.

Aluminum foil is a good conductor of heat, so it is not a good material for keeping heat away from the ice cube. Air is a poor conductor of heat, so materials with air pockets are good at insulating the ice cube from heat.

Good Ice Keepers would also make good winter coats. In that case, they trap heat inside instead of keeping it out.

How can your child make a better Ice Keeper? Consider adding more material, using a bigger cup, or using a combination of materials.

© Scholastic Inc.

Wind Power

Background

We harness the power of wind as a renewable energy source. You may have seen **wind turbines**—machines that turn wind energy into electrical energy (see photo below). Wind turns a turbine's blades, which spins a generator inside the turbine to produce electricity.

Hands-On Hints

Investigation 1

Blowing on the stars and triangles (at the tips of the "wings") will work best to turn the Spin Wheel. Blowing on the circles and straight down may cause it to turn some. Blowing straight up will send the Spin Wheel flying off the pencil.

Investigation 2

For a Super Spinner Cone to stay on the pencil, its weight must be evenly balanced. Though it is mentioned in the instructions, you may want to talk about balance with your child ahead of time. Simply put, *balance* is when something stays in a stable position. It is related to *center of gravity*—the point around which all of an object's weight is evenly balanced. When the spinner's center of gravity is at the same point as its point of support (the tip of the pencil), it will be balanced. The easiest way to keep a spinner balanced is to make sure each side of the spinner carries about the same weight. So if your child adds a feather on one side, for example, he or she should add another feather or some other object that weighs about the same on the opposite side.

Provide a wide variety of lightweight materials, such as pipe cleaners, straws, and feathers, which are available at craft stores. (Don't use feathers found outside.) Encourage your child to think about how to design the Super Spinner Cone to "catch the wind." If you make the Super Spinner Cone using card stock, your child can use heavier materials.

Forces

Background

"Soapbox" is the term commonly used for a four-wheeled, motorless vehicle that is raced on a downhill road. Another name for it is "gravity racer." Typically, a heavier cart will do better than a lighter one. However, lighter wheels tend to accelerate faster than heavier wheels. The ramp or track also affects how a cart will perform. A cart that goes fast down one track may not do as well on a different track. Racers take all of these and other factors into account when they build and design their soapbox racers.

Hands-On Hints

Investigation 1

Use small plastic bottles, like individual serving water bottles or 20-oz. soda bottles. If possible, use matching pairs of empty and full bottles so it's clear that you're changing just one variable.

Your child should be able to feel that the full bottle gives a bigger push, with more force (Steps 3 and 4). Both rolling bottles will give the box a push away from the ramp. The heavier full bottle will push the box farther (Steps 5 and 6).

Investigation 2

Remind your child to test one change at a time. This way, he or she will know which change led to a difference in performance.

For experimental purposes, provide additional blocks (for making a taller ramp), different-sized boxes, and different-sized bottles. If you have something that can serve as a longer ramp, make that available as well. If you have multiple floor surfaces (like carpet vs. tile), you might suggest that your child think about changing the ramp's location.

Your child may find that the following changes increase the distance the box moves: a steeper ramp (unless it gets too steep), a longer ramp, a lighter box, a smaller box (assuming it's not so small the bottle rolls over it), a heavier bottle, a bigger bottle (assuming it doesn't roll over the box), or a smoother floor surface.

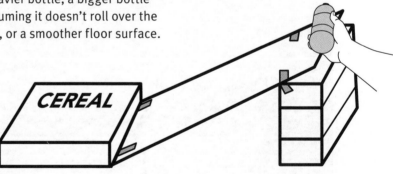